C000077217

IMAGES
of America

BUILDING THE
BLUE RIDGE PARKWAY

VIADUCT CONSTRUCTION, C. 1938. Round Meadow Viaduct was constructed in November 1938. Not well-known, it is one of three viaducts that can be found on the Blue Ridge Parkway. Concrete with steel reinforcements were used to build this large bridge. Viaducts allow for the road to stay level rather than dipping deep into ravines. (Courtesy of the Blue Ridge Parkway Archives.)

ON THE COVER: The cover photograph shows a large shovel used for Albert Brothers Construction out of Salem, Virginia, sometime in the 1930s. Charles Albert and Albert Brothers Construction first appeared in the Roanoke, Virginia, city directory in 1939. The list of officers was the same till 1955. Albert Brothers Construction listed Lincoln J. Messimer as president and treasurer, George D. Abernathy as vice president, and Roland H. Clark as secretary. Their address was 1102 Tennessee Street in Salem, Virginia. In 1960, Albert Brothers Contractors listed Lincoln J. Messimer as president and treasurer, George D. Abernathy as vice president, and Kathy A. Burk as secretary. In 1974, Albert Brothers Contractors, Inc., listed L. J. Missimer as chairman of the board (note spelling change). C. D. Missimer was listed as president and treasurer. R. E. Buchanan, C. R. Faries, and M. E. Goad were all listed as vice presidents; Elaine Vaughn was listed as secretary. It is unknown exactly how many contracts they acquired with the National Park Service. Photographs indicate that they had projects in Virginia and in North Carolina. (North Carolina State Archives Photograph Collection.)

IMAGES
of America

BUILDING THE
BLUE RIDGE PARKWAY

Karen J. Hall and
FRIENDS of the Blue Ridge Parkway, Inc.

ARCADIA
PUBLISHING

Copyright © 2007 by Karen J. Hall and FRIENDS of the Blue Ridge Parkway, Inc.
ISBN 978-0-7385-5287-3

Published by Arcadia Publishing
Charleston, South Carolina

Printed in the United States of America

Library of Congress Catalog Card Number: 2007926284

For all general information contact Arcadia Publishing at:
Telephone 843-853-2070
Fax 843-853-0044
E-mail sales@arcadiapublishing.com
For customer service and orders:
Toll-Free 1-888-313-2665

Visit us on the Internet at www.arcadiapublishing.com

CONTENTS

ACKNOWLEDGMENTS

Kudos is extended to FRIENDS of the Blue Ridge Parkway for their full support and for all of the wonderful work they are doing along the Blue Ridge Parkway. Without their hard work and commitment, we as a nation might lose our national treasure. I would like to recognize all of the former conservation workers and contractors for all of the hard work they put into building the parkway. What a legacy they have left.

In partnership with the National Park Service as well as with other organizations, businesses, and agencies, the staff and membership of FRIENDS have made it their mission to protect the parkway and raise public awareness of this national treasure so that future generations will continue to experience the unique journey it offers.

FRIENDS wishes to thank Karen J. Hall, author of this book and the Blue Ridge Parkway's Postcard History Series book, for her major contribution to FRIENDS through sharing coauthorship of these books with FRIENDS of the Blue Ridge Parkway. Karen is not only a member of FRIENDS but is passionate about the Blue Ridge Parkway and making a difference for future generations. Contributions such as Karen's are priceless to the future of America's Most Scenic Drive.

FRIENDS of the Blue Ridge Parkway, Inc., a 501(c)(3) nonprofit corporation, organized and exists under the laws of the State of North Carolina and the Commonwealth of Virginia. Consider leaving a Blue Ridge Parkway Legacy to support the Blue Ridge Parkway's future by remembering FRIENDS in your will.

For membership information or donations, contact:
FRIENDS of the Blue Ridge Parkway, Inc.
Post Office Box 20986
Roanoke, VA 24018
1-800-288-PARK (7275)
www.BlueRidgeFriends.org

INTRODUCTION

[The Blue Ridge Parkway] has but one reason for existence, which is to please
by revealing the charm and interest of the native American countryside.

–Stanley Abbott

The Blue Ridge Parkway's resident architect, Stanley Abbott, felt the basic premise for the parkway's existence was to reveal the "charm and interest of the native American countryside." It was up to Abbott to determine the design principles for the construction of the Blue Ridge Parkway. He had historic examples of other parks to consider, such as the Westchester Parks Commission and the Shenandoah National Park, but Abbott chose to design the Blue Ridge Parkway in a unique style. Abbott and his colleagues developed the following principles to guide the construction and future operations of what is called "America's Most Scenic Drive." The principles that were applied to create the Blue Ridge Parkway today are:

The horizon would be the boundary as the parkway traverses a protective corridor.

The structures to be associated with the Blue Ridge Parkway were to be simplistic and informal to harmonize with the natural environment.

All elements of the parkway would be related to one another, supporting one another—unified.

Variety would be the spice of the parkway, mixing mountain vistas, rolling hillsides, and dense woodlands.

The drive should be easy and safe as not to distract from the views.

The road should be a marriage of the cultural landscape with the natural landscape.

The roadside landscape would accept the responsibility to preserve and interpret the cultural history, "revealing the charm and interest of the native countryside."

Finally, the road would preserve the "whole scenic picture" for the visitor to enjoy recreation, hikes, overlooks, and lodging, according to Abbott, "like beads on a string"—the rare gems in the necklace.

The creation of our natural treasure, the Blue Ridge Parkway, was a great feat accomplished by many people orchestrating a vision that exists today for our enjoyment. Stan Abbott expressed through the following quote that it was far more than man creating a parkway, but it was a mission of soul and spirit: "I can't imagine a more creative job than locating the Blue Ridge Parkway, because you worked with a ten league canvas and a brush of comet's tail. Moss and lichens collecting on the shake roof of a Mabry Mill measure against huge panoramas that look out forever."

Personnel from the Civilian Conservation Corps (CCC) shared much in common with FRIENDS of the Blue Ridge Parkway. The CCC boys did landscaping, park benches, view sheds, and visitor centers. Today FRIENDS restores the view shed and landscaping. In addition, our volunteers construct trails and provide benches such as those at the Parkway's Blue Ridge Music Center. Volunteers now staff the visitor centers and assist with cleanup after hurricanes and storm damage at visitor centers, trails, and roadways.

Today the Blue Ridge Parkway is a special place—a 469-mile road that climbs the ridgelines and peaks of the Appalachian Mountains, every year carrying 20 million visitors between Shenandoah National Park and the Great Smoky Mountains.

The Blue Ridge Parkway is the most visited park nationally. The first principle—the horizon was the boundary as the parkway traverses a protective corridor—is changing. This is the very reason Dr. Harley Jolley has termed it an "endangered species." Residential and commercial growth, pollution, and non-native predatory insects are taking their toll on the parkway's views, trees, and wildlife. It is not going unnoticed.

As of 2007, membership in the Blue Ridge Parkway's 501(c)(3) nonprofit organization, FRIENDS of the Blue Ridge Parkway, Inc., has topped 7,500. FRIENDS inspires the parkway visitor to join one of the fastest-growing "friends groups" nationally to preserve and protect the Blue Ridge Parkway. With growing support from our readers and members, FRIENDS has been able serve the Blue Ridge Parkway since 1989.

Not only are there over 7,500 individuals who recognize the historic and natural value of the Blue Ridge Parkway—and the risk to its survival—but FRIENDS spearheads the volunteer efforts for the Blue Ridge Parkway. In 2006, FRIENDS increased the hours of volunteer service to the parkway by 30 percent.

The parkway is unusual among national parks in that it is linear, passing through two states (Virginia and North Carolina) and 29 counties, touching human and natural communities along the way. The parkway, called America's Most Scenic Drive, is within one day's drive of more than half of the population of the United States.

The parkway is a rich tapestry, weaving natural beauty and history with human cultural heritage. It is America's living rural life museum, where basket-weavers, traditional Appalachian musicians, blacksmiths, and quilters bring the past to the present. Historic sites such as gristmills, bridges, and one-room schoolhouses exist along the parkway, and over 350 miles of trails crisscross its length, connecting with the Appalachian Trail and even older pathways that wind through forests of hardwoods, evergreens, and the endangered hemlock.

All special places of great beauty are affected by time, and the Blue Ridge Parkway is no exception. Residential and commercial development is compromising its views and wildlife habitats; pollution is fogging its clear air; and predators are destroying its trees. Dr. Harley Jolley, historian and national authority on the Blue Ridge Parkway, states, "The Blue Ridge Parkway, America's living rural life museum, is becoming an endangered species."

Recognizing these threats, FRIENDS has pledged to help preserve, protect, and promote the outstanding natural beauty, ecological vitality, and cultural distinctiveness of the Blue Ridge Parkway and its surrounding scenic landscape, preserving this national treasure for future generations.

Founded in 1989 by the Blue Ridge Parkway superintendent, FRIENDS was organized as a nonprofit, membership-based organization to provide a link between parkway visitors and the parkway experience. It was believed that the organization and its members could be a catalyst for ensuring the preservation, conservation, and enhancement of the parkway, which is not only a scenic asset but an economic one, bringing millions of visitors and tourists to Virginia and North Carolina every year.

A growing membership base reflects FRIENDS' vitality and grassroots success. Many of the organization's key projects enlist the labor and energy, as well as the financial support, of members to accomplish its goals. Here are our key projects:

SAVE PARKWAY VIEWS. Both FRIENDS and the U.S. Park Service feel that the parkway's greatest threat today is encroaching development that impacts the visual experience. In 2003, Scenic America designated a 28-mile section of the parkway as a Last Chance Landscape, and that phrase can be loosely applied to all 469 miles as communities expand and industry grows and natural, pristine places grow smaller and more rare.

The work of FRIENDS includes raising money, soliciting donations, and involving community volunteers in planting trees along the parkway to buffer undesirable views. These view sheds are landscaped to restore visitors' visual experiences along the parkway, to reestablish wildlife habitats along the parkway corridor, and to rebuild an ecological buffer against development. In recent years, FRIENDS has received awards for its restoration of the parkway views. In 2004, Scenic Virginia presented FRIENDS with the second Annual Scenic Award for best preservation of a scenic view shed.

In 2005, the oldest preservation group in the nation, Association for the Preservation of Virginia Antiquities, presented FRIENDS with the first-ever Founder's Award for their efforts related to the preservation of this tremendous asset.

A primary concern of visitors to the Blue Ridge Parkway is the loss of natural views. As the only authorized agency to restore parkway views, FRIENDS has identified over 50 views along the 469-mile parkway that need immediate attention. Each view restoration has a price tag of over $20,000. It is essential that FRIENDS continues the Stan Abbott legacy to re-create the Blue Ridge Parkway to reflect the original design principles.

Through our Save the Views program, hundreds of mature trees and seedlings are being planted to re-create the design elements of the original parkway design strategy.

In addition, our Kids Empowering Kids program has children of all ages planting the trees. Youths also mentor younger children to plant trees and seedlings, gain environmental awareness, and share in the memories, saying, "This is my tree—and over the years to come, I will watch it grow."

Volunteers and communities along the parkway are joining hands to support each project. Mature trees are planted in the fall and seedlings in the spring, with our last planting achieving a 98-percent survival rate.

THE VOLUNTEERS-IN-PARKS (VIP) PROGRAM: ENRICHING OUR HERITAGE. This program is a vital part of the National Park System. Volunteers-In-Parks are Very Important People (VIPs). In fiscal year 2006, FRIENDS increased the volunteer hours of service to the parkway by over 30 percent.

As corporate giving and federal funding decrease, National Park Service staff and resources become more limited. Volunteers play a critical role in filling the gaps by providing innumerable services, which include assisting visitors in welcome centers, maintaining trails, staffing campgrounds, and providing historical interpretation and cultural demonstrations.

Volunteers' contributions to the experiences of parkway visitors allow the parkway to continue to educate contemporary travelers about the significance of the region's historical and cultural rural past.

The VIP Program, administered by FRIENDS, recruits and instructs volunteers from communities, universities, and businesses along the parkway's corridor and raises funds for volunteer programs and recognition events.

There are many dream jobs for the much-needed volunteers along the 469 miles of the Blue Ridge Parkway. Due to the major shortfall of over $600 million impacting the National Park Service, your support is needed more than ever!

The following are examples of the impact upon the parkway: campgrounds are closed because of lack of volunteer hosts, visitor centers cannot open without volunteers, visitors' experiences are diminished if there are no cultural heritage demonstrations. The list is endless. Volunteers are the lifeblood of the Blue Ridge Parkway.

Staff shortages, the "endangered ranger," and lack of park funds all diminish the visitor's experience and the ability of the Park Service to preserve America's heritage.

Our Volunteers: Enriching our Heritage Program helps the bog turtle, a federal endangered species that lives along the wetlands adjacent to the Blue Ridge Parkway. FRIENDS provides transmitters that track the turtles, allowing biologists to learn what parts of the wetlands the turtles use, whether they also move between other wetlands nearby, how they deal with the traffic on motor roads, and where they spend the winter. This is just one way volunteers and FRIENDS make a difference.

TRAILS FOREVER PROGRAM. This program seeks to maintain Blue Ridge Parkway trails in order to enhance forever our access to over 350 miles of trails along our parkway.

FRIENDS recruits for and administers the Trails Forever Program for the National Park Service/Blue Ridge Parkway in a similar approach to the VIP program—asking people of all ages and walks of life to work together to build and maintain trails on parkway land.

Community-based action is at the heart of FRIENDS' volunteer-based organization. Adopt-A-Trail and other projects like it empower community-based groups along the Blue Ridge Parkway corridor to collaborate in preserving a treasure in their own backyards.

This vast trail system is an extensive access network connecting millions of people to these natural landscapes.

The objective of Trails Forever is to improve the quality of the Blue Ridge Parkway trails, enhance the experiences of the park user, and engage the corridor communities in sustaining the park's trail system forever. FRIENDS is working with the park service to expand and build a first-class trail system, but we can only do this with the help of volunteers.

FRIENDS recruits Trail Groups to construct trails, build bridges, and provide landscaping to improve the quality of our trails.

Our goal also is to have Trail Ambassadors that can help visitors learn more about the park's natural resources as they hike the trail.

PRESERVE THE HEMLOCKS. Help us defeat the woolly adelgids that threaten our unique and spectacular hemlock forest. With your support, our magnificent hemlocks will continue to thrive for years to come—and be enjoyed by future generations.

The hemlock woolly adelgid is native to Asia, where it is not a problem to local hemlocks. In 1950, it was introduced to the eastern part of the United States, where it does not have natural enemies in our environment—and is therefore a deadly threat.

Recently, this small, relentless, deadly non-native pest is showing up in places where it has never been seen before—on majestic host trees that were not expected to be hit for several years. Experts say the region faces an ecological tragedy that parallels the chestnut blight.

"It was a dark, dark day," says Parkway Resource Management Specialist Lillian McElrath, of her discovery of adelgids "pretty much everywhere" at Linville Falls: in the picnic area, along the river, around the visitor center, on trails to overlooks, and in Linville Gorge. "I think we'll be seeing tree death at Linville Falls within the next two years, she says. "Once the adelgid hits, it's a pretty quick thing."

For years, FRIENDS has made donations to research facilities to address this issue, but funds are greatly needed to make a difference before our hemlocks are destroyed.

In a time when the world is becoming increasingly uncertain and when our view of the future is less than clear, the Blue Ridge Parkway stretches along mountains that are ageless, that have been here since long before we lived, and that will be here for centuries to follow. Through the support of FRIENDS, future generations will experience Stanley Abbott's basic premise for the Blue Ridge Parkway—the Blue Ridge Parkway "has but one reason for existence, which is to please by revealing the charm and interest of the native American countryside."

–Susan Jackson Mills, Ph.D.

One

BEFORE CONSTRUCTION

Protecting a beautiful view and creating economic growth was the initial goal behind the construction of Skyline Drive and Shenandoah National Park.

By 1700, nearly 68,000 inhabitants lived in the Blue Ridge Mountains of Virginia. First documentation of exploration materialized in 1716, when Gov. Alexander Spotswood traveled up the Rappahannock River. This area became officially recognized and land grants were issued encouraging Virginians to make this area their home.

Early American roads were nothing more than trails that had been widened for wagons. Very few had gravel. One of the most famous early roads was the Great Wagon Road. Hordes of early German and Scotch-Irish settlers used what became known as the Great Wagon Road to move from Pennsylvania southward through the Shenandoah Valley, through Virginia and the Carolinas, to Georgia, a distance of about 800 miles. The mountain ranges to the west of the valley are the Alleghenies, and the ones to the east constitute the Blue Ridge chain.

Once invented, the automobile changed the country. Clubs called auto clubs developed and began asking for roads into our national parks. The first park with roads was Yellowstone. Then came the request for the Blue Ridge Parkway.

This chapter contains photographs of mountain folk, farmland before it was acquired, early mountain roads, and Civilian Conservation Corps camps and enrollees.

The CCC boys did not construct the roadbed of the parkway, as you will see in chapter two; contractors did most of the roadbed work. They graded slopes and did most of the landscaping by planting thousands of trees and shrubs and acres of grass to landscape on both sides of the road. They built guardrails, guard walls, and overlooks.

EARLY ROAD IN 1909. Roads in the early 1900s were treacherous with deep ruts. Each road was maintained by hand, like this one in the Western Mountains of North Carolina. In 1908, Henry Ford introduced the first Model T, changing the way Americans dreamed of traveling forever. By 1916, the Federal Aid Road Act had been passed, funneling funds to states for road building. (North Carolina State Archives Photograph Collection.)

BURKE COUNTY OX AND CART IN 1906. Transportation for farmers consisted of oxen in the 1800s and early 1900s. They are powerful animals and could pull heavy loads, even in the mud. This photograph shows a Burke County, North Carolina, farm family heading to town in their family wagon, pulled by a team of oxen. (North Carolina State Archives Photograph Collection.)

BOONE PICNIC IN THE EARLY 1900s. Traveling to the mountains for a Sunday picnic was a tradition long before the construction of the Blue Ridge Parkway. This photograph shows a group of young people that have hiked up to the top of one of the high peaks in Boone, North Carolina. (Courtesy of Historical Boone.)

CHAIN GANG IN 1930. This photograph, taken in the early 1900s, shows a chain gang doing road maintenance. Chain gangs performed menial tasks like chipping rocks for roadbeds. It is doubtful that any chain gangs worked on the construction of the Blue Ridge Parkway. By 1955, chain gangs were phased out in the United States. (North Carolina State Archives Photograph Collection.)

OLD MOUNTAIN ROAD IN 1913. This 1913 photograph of an automobile turning the curve was taken on the road in the Blue Ridge Mountains. Mr. Pratt's road was surveyed and constructed near Linville, North Carolina. (North Carolina State Archives Photograph Collection.)

MRS. MACE ON FEBRUARY 14, 1913. A Sunday afternoon walk or picnic surely raised the curiosity of locals wanting to know what this new Blue Ridge Highway would look like. On February 14, 1913, this photograph was snapped of an older couple, Mrs. Mace and a friend, checking out the road in Western North Carolina. (North Carolina State Archives Photograph Collection.)

RUSS NICHOLSON IN OCTOBER 1935. Landowners like Russ Nicholson of the Shenandoah Valley either had to sell or donate their land taken for the path of the Skyline Drive and the Blue Ridge Parkway. Russ lived in Nicholson Hollow, where his ancestors had settled 200 years earlier. Deeds and rights-of-way had to be acquired before construction could begin. (Courtesy of the Library of Congress.)

NETHERS POST OFFICE IN OCTOBER 1935. Photographer Arthur Rothstein took photographs of the resettlement folks of the Shenandoah Valley for the Farm Security Administration. The story of the Blue Ridge Parkway construction cannot be told without telling the story of the people. Limited photographs are available of the people who lived along the mountain tops that became the Blue Ridge Parkway; thus, photographs from the Shenandoah Skyline Drive have been used here to illustrate an important piece of history. This is one of his photographs showing the outside porch of the Nethers, Virginia, post office. The Farm Security Administration (FSA) was the Resettlement Administration in 1935 as part of the New Deal. It was an experiment in collectivizing agriculture—that is, in bringing farmers together to work on large government-owned farms using modern techniques under the supervision of experts. The program failed because the farmers wanted ownership, and the agency was transformed into a program to help them buy farms. The FSA bought out small farms that were not economically sound and combined them with about 33 other farms in a co-op type of situation where the farmers lived in a community and worked together. (Courtesy of the Library of Congress.)

INTERIOR POSTMASTER'S OFFICE IN OCTOBER 1935. This photograph shows the interior of the postmaster's home-based office in Nethers, Virginia. The postmaster played a very important role in rural America. He faced the challenge of delivering the mail over all types of terrain and weather conditions. (Courtesy of the Library of Congress.)

WHEAT THRESHING ABOUT 1920. Farming in Boone, North Carolina, in the early 1900s required a lot of actual horse power and man power. Before the tractor became an affordable piece of equipment, farmers had to do a lot of the work by hand. This is the type of farm that the Blue Ridge Parkway traversed after construction. (Courtesy of Historical Boone.)

NATURAL BRIDGE AROUND 1910. Originally, the Natural Bridge was supposed to be part of the Blue Ridge Parkway. As it stands, the parkway was shifted several miles east, a short distance from the natural wonder. Today this spot boasts a motel, visitor center, and wax museum. (Courtesy of Karen Hall.)

AN OLD MILL. This photograph of Estelle O. McDaniel near an old mill was taken in the early 1950s during a Sunday outing. Historically gristmills contained rotating stones powered by water or by wind. Mabry Mill at milepost 176.1 on the Blue Ridge Parkway is powered by water. (Courtesy of the Hardy family.)

MOUNTAIN MAN AND DONKEY IN EARLY 1900. Donkeys and pack mules were important for hauling supplies in the 1800s. They were cheaper than horses and stronger for hauling loads. They also need less food than horses. Donkeys have developed very loud voices, which can be heard over several miles. They can defend themselves with a powerful kick of their hind legs. (Courtesy of the Blue Ridge Parkway Archives.)

CCC MOVING IN TO DOUGHTON PARK IN JUNE 1938. The Civilian Conservation Corps were part of the New Deal, arranged by Franklin D. Roosevelt for economic relief during the Depression. This is one of the first camps set up for construction of the Blue Ridge Parkway. It was in Bluff Park, which is now Doughton Park, milepost 238.6. Landscaping was their main function. (Courtesy of the Blue Ridge Parkway Archives.)

SAM LOWE IN 1940. Sam Lowe was an enrollee at the Doughton Park CCC camp in Laurel Springs, North Carolina. You can see one of the barracks behind him. He would have learned work skills to take to the work world when the CCC was disbanded. The men worked 8 hours per day for a total of 40 per week. (Courtesy of the Blue Ridge Parkway Archives.)

CCC COMPANY OF DOUGHTON PARK IN 1940. In this photograph, one can see that the camp is more established—with buildings instead of tents. The buildings would have included living barracks, a library, mess hall, storage facilities for equipment, and a recreational facility. (Courtesy of the Blue Ridge Parkway Archives.)

ENROLLEES OF DOUGHTON PARK IN 1940. These young gentlemen were enrollees at the Doughton Park CCC camp in Laurel Springs, North Carolina. From left to right, they are identified as Dale Shepherd, unidentified, Travis Owens, and Arthur Phipps. They are standing at a directional sign showing the way to the picnic area or the trail to Wildcat Rock. (Courtesy of the Blue Ridge Parkway Archives.)

PETE KULYNYCH IN 1940. Pete Kulynych became administrative assistant with Lowe's Hardware in North Wilkesboro, North Carolina, after leaving the CCC. The purpose of the CCC was to hire local boys and train them in usable skills. Most of their work projects included landscaping, but they also built comfort stations, guttering, and slope grades. (Courtesy of the Blue Ridge Parkway Archives.)

CCC ENROLLEES IN 1940. From left to right, these young men are Bert Richardson and Robert Wagoner. They were stationed at the Doughton Park CCC camp in Laurel Springs, North Carolina. Richardson was a native of the Blue Ridge Mountains. Local boys had to fill out an application and go through an interview process to be employed by the CCC. (Courtesy of the Blue Ridge Parkway Archives.)

CCC Camp at Doughton Park in 1940. This is a great example of how the CCC camps developed into a small community for the young men that enrolled with them. At least 17 buildings can be counted in this photograph. Buildings were premanufactured at a plant in Virginia so that they could be disassembled and reassembled if needed. After the end of the CCC, buildings were dismantled and the lumber was recycled. (Courtesy of the Blue Ridge Parkway Archives.)

Ed Carico in 1940. Ed Carico is thought to have been one of the foremen at the Doughton Park CCC camp. He is standing next to a vintage truck that would have been his means to travel from project to project directing the men. (Courtesy of the Blue Ridge Parkway Archives.)

CURTIS CREEK CAMP ON APRIL 15, 1936. Curtis Creek Camp, located near Old Fort, North Carolina, was a Forestry Service Camp in McDowell County. Most of the CCC buildings were treated with creosote, a chemical that preserves the wood from insects and weather. The men from this camp helped the CCC camps on the parkway with tree planting and forest-fire management. (Courtesy of the Blue Ridge Parkway Archives.)

ROCK CASTLE GORGE CAMP ON MAY 23, 1938. Located within 3.2 miles from the Rocky Knob picnic area in Floyd County, Virginia, the Rock Castle Gorge Camp had enrollees that would have built the trails, comfort stations, and picnic areas. The overlook for Rocky Knob would have been built near this camp. All that is left of this camp today are the building foundations. (Courtesy of the Blue Ridge Parkway Archives.)

ROCK CASTLE GORGE ON MAY 23, 1938. The tents indicate that the camp was very young in development. They were started with heavy army tents until facilities could be built to house the men. Grass had to be mowed and brush had to be cleared to keep the camp inhabitable. (Courtesy of the Blue Ridge Parkway Archives.)

BLACK CAMP GAP ON AUGUST 20, 1935. Located at milepost 3.3, the Black Camp Gap boasted at least 21 tents when first built. The tents were bigger than the Model T sitting to the right. With the flaps open it would appear that it was pretty hot in August 1935. A cool summer breeze probably brought much needed relief. (Courtesy of the Blue Ridge Parkway Archives.)

BARRACK

CAMP BARRACKS IN NOVEMBER 1948. Neatly made beds lined the barracks of the Peaks of Otter Camp. Work from the forestry division at Kelso, Virginia, was transferred to the Peaks of Otter in 1934. The men were involved in remediating fire hazards and selective landscaping. They also constructed the James River CCC camp in 1941. (Courtesy of the Blue Ridge Parkway Archives.)

MESS HALL

PEAKS OF OTTER MESS HALL IN 1948. Nourishing hot meals were provided for dinner every day and sometimes for breakfast. Lunch was usually a sandwich with a piece of fruit or pie for dessert. This mess hall has the tables set with coffee cups, plates, and napkins. Packaged buns are stacked on the table, waiting on the next meal. (Courtesy of the Blue Ridge Parkway Archives.)

HAND DRILLING IN 1935. Pneumatic drills, powered by air compressors, are used in open excavation to drill holes into solid rock. A steel bit and shank is pounded and turned by air-power to bore the hole. The earliest pneumatic drills were hand-held or mounted on stands, singly or in groups. Known as "sinker drills," hand-held drills are still used today in specialized applications. (Courtesy of the Blue Ridge Parkway Archives.)

CAMP NIRA IN THE SHENANDOAH VALLEY IN 1933. The first CCC camp was built in the Shenandoah National Park (not the Blue Ridge Parkway). Later some of the trained men became park rangers for the Blue Ridge Parkway. (Courtesy of the Shenandoah National Park.)

GROTTOES CPS CAMP IN 1941. Camp Grottoes was the fifth Civilian Public Service (CPS) camp built in the Shenandoah National Park. It was located in Virginia. The camp was active during World War II and up until 1947. (Courtesy of the Shenandoah National Park.)

ROCK CRUSHER IN AUGUST 1936. Before construction could get too far under way, rock crushers had to be set up at various locations to provide the necessary crushed gravel to make the roadbed. This crusher operation was located near milepost 150.5 in Virginia. When possible, they recycled local rock into the roadbed. (Courtesy of the Blue Ridge Parkway Archives.)

Two

EARLY CONSTRUCTION

Creating a scenic highway took hard work. Many contractors were involved in the creation of the Blue Ridge Parkway.

Contractors constructed most of the roadway. They had to bid on the work by sections. The first contract for work on the Blue Ridge Parkway, awarded to Nello Teer Construction Company of Durham, North Carolina, in the height of the Great Depression, was for over $316,000. Some of the other contractors were the following: Albert Brothers Construction, Inc., of Salem, Virginia; M. E. Gilioz Company of Monett, Missouri; Chandler Brothers, Inc., of Virgilina, Virginia; Ralph E. Mills Company of Frankfort, Kentucky; W. H. Armentrout of Harrisonburg, Virginia; Keeley Construction Company of Clarksburg, West Virginia; Waugh Brothers of Fayetteville, West Virginia; and the U.S. Department of the Interior.

Bituminous surfacing of the roadway was separately contracted and was awarded to Corson and Gruman Company of Washington, D.C.; Southern Asphalt Company of Richmond, Virginia; or Barrett Paving Company of Harrisonburg, Virginia.

One of the most well-known construction organizations was the Civilian Conservation Corps (CCC). The CCC was a work relief program for young men established in March 1933 during Pres. Franklin D. Roosevelt's first 100 days in office. Being part of the New Deal, it was designed to combat the poverty and unemployment of the Great Depression. The young men did heavy construction work (like roadwork) and did not receive any professional training.

Another similar program was the Works Progress Administration (later Work Projects Administration, or WPA). The WPA was created in May 1935 by presidential order. It was the largest and most comprehensive New Deal agency.

All of these programs were labeled "Alphabet Soup" by detractors. Most have been dissolved but a few remain today. The Tennessee Valley Authority (TVA) and the Federal Deposit Insurance Corporation (FDIC) are two that remain in existence today.

CUMBERLAND KNOB VISITOR CENTER IN 1935. Construction began at the Cumberland Knob area in 1935. On September 21, 1939, the Cumberland Knob Visitor Center was opened to the public. This photograph shows the construction of the visitor center, which still exists. Services have been relocated to the Blue Ridge Music Center a short distance away. (Courtesy of the Library of Congress.)

COMPLETED VISITOR CENTER IN SEPTEMBER 1939. Hand-hewn logs were used to construct the comfort station and visitor center for Cumberland Knob. It is located at milepost 217.5 along the Blue Ridge Parkway, just over the North Carolina state line from Virginia. The Cumberland Knob trailhead can be accessed here. On 1,000 acres, one can have a picnic or take a hike on the trail. (Courtesy of the Library of Congress.)

BLASTING ROCK IN 1940. This photograph looks north on the parkway somewhere between mileposts 348 and 360. Rock blasted from this site would have been recycled either by crushing it and using it in the roadbed or creating sloped banking. Danger accompanied the dynamite users. This project was completed in 1941. (Courtesy of the Blue Ridge Parkway Archives.)

CUMBERLAND KNOB TRAIL SHELTER IN SEPTEMBER 1939. A short walk from the visitor center is the trail shelter. Made of rough-hewn logs, it provides shade and shelter from the weather along with a very nice fireplace. The foundation consists of river rocks that were recovered from the nearby creek. This spot bears a spectacular view of Gully Creek Gorge. (Courtesy of the Library of Congress.)

SHELTER WITH FIREPLACE IN SEPTEMBER 1939. The trail shelter at Cumberland Knob houses a wonderful fireplace for roasting marshmallows and relaxing. It will comfortably hold 20 people. Unfortunately visitors have written graffiti all over the walls of this wonderful little spot. This photograph shows the original condition of the shelter. (Courtesy of the Library of Congress.)

HAND LEVELING ON JULY 22, 1936. The first project of the Blue Ridge Parkway was section 2-A, near the North Carolina state line. Hand leveling was done before the asphalt was applied to the roadbed to make sure that all measurements were accurate. (Courtesy of the Blue Ridge Parkway Archives.)

WAGON DRILL IN ACTION IN SEPTEMBER 1936. This wagon drill was photographed near milepost 244.9 in September 1936. This was the third section of construction of the Blue Ridge Parkway near the Bluffs, now called Doughton Park. The drills made holes for dynamite, and rock was blasted so that it could be hauled away to make room for the road. This area has large outcroppings of native stone. (Courtesy of the Blue Ridge Parkway Archives.)

BRINEGAR CABIN PARKING LOT IN 1942. The Brinegar Cabin parking area, located at milepost 238.5, was constructed about 1942 by the U.S. Department of the Interior. An early bucket truck and dump truck are seen here loading up blasted rock. Once finished, this parking lot will hold approximately 30 cars. (Courtesy of the Blue Ridge Parkway Archives.)

MILEPOST 235 CONSTRUCTION IN SEPTEMBER 1936. Construction between North Carolina Highway 18 and the Brinegar Cabin began in March 1936 and was completed in December 1937. This early track tractor was great for pulling heavy loads in uneven terrain. Another one can be seen dumping to the side. A truck with a blade can be seen grading the road. (Courtesy of the Blue Ridge Parkway Archives.)

FILLING LOW AREA IN NOVEMBER 1936. Fill dirt had to be hauled in to fill this low area north of the Brinegar Cabin area of the Blue Ridge Parkway at milepost 221. Wetlands were encountered throughout the construction. One of the features of the Blue Ridge region of the Appalachian Mountain range is the many streams and wetlands providing lush vegetation. (Courtesy of the Blue Ridge Parkway Archives.)

ROCK SLIDE AT BLUFF PARK IN THE LATE 1930S. Rock slides are very common in the Blue Ridge Mountains. This road crew is seen with shovels beginning to remove the debris left by the large rock slide near Bluff Park (now Doughton Park). A single white line divided the road. Double yellow lines weren't used until the 1950s. (Courtesy of the Blue Ridge Parkway Archives.)

MAHOGANY ROCK OVERLOOK CONSTRUCTION IN AUGUST 1936. Located two miles north of Brinegar Cabin, the Mahogany Overlook gives a great view of Sparta, North Carolina, the county seat of Alleghany County. This photograph shows a man using a hand drill and a large bucket excavator from Albert Brothers Construction from Salem, Virginia. The excavator, sometimes called a 360-degree excavator, consists of a bucket and cab placed on a pivot atop an undercarriage with either tracks or wheels. Independent contractors were hired to do most of the road construction along the parkway. They were required to bid on the project, and the lowest bidder was selected. Albert Brothers Construction also worked on Skyline Drive completing several sections. They completed the sections between Hogback Mountain and Thorton Gap, Black Rock Gap and Jarman Gap, and many others. (Courtesy of the Blue Ridge Parkway Archives.)

OVERPASS CONSTRUCTION ON SEPTEMBER 9, 1938. When plans were being made for the Blue Ridge Parkway, it was decided that all of the structures, including overpasses and bridges, were to be faced with native rock. Here one can see how the project developed. The skills of the mason were in high demand. (Courtesy of the Blue Ridge Parkway Archives.)

ROARING GAP OVERPASS STRUCTURE IN 1938. The bridge over Highway 221 can be seen here in its construction phase on September 8, 1938. The bridges were started with a semicircle steel base and then reinforced with steel and rock facing. Contractors and not the Civilian Conservation Corps would have built them. (Courtesy of the Blue Ridge Parkway Archives.)

FOOTPRINTS ON THE BLUE RIDGE PARKWAY ON OCTOBER 20, 1936. Footprints in the dirt lead upward where a guardrail is badly needed on this section of the Blue Ridge Parkway, milepost 262.3. This looks over the Yadkin Valley for many miles. This portion of the parkway was under construction in September 1936 and completed in 1941. Located in Alleghany County, North Carolina, views are spectacular and the sky is very blue on a nice clear day. The closest CCC camp would have been located at Doughton Park. Named for Congressman Robert L. Doughton, a loyal supporter and neighbor of the parkway, this is an ideal location to see deer. Formerly known as the Bluffs because of its high, open meadowlands, this 7,000-acre park has picnic areas, trailer sites, a campground, and comfort stations, as well as 30 miles of hiking trails over bluegrass bluffs. (Courtesy of the Blue Ridge Parkway Archives.)

CUTTING AND GRADING ON NOVEMBER 26, 1935. This section of grading and cutting of the road bed was part of the first contract for the Blue Ridge Parkway. Section 2-A contained a 12.695-mile length and was near present-day milepost 221.5. The terrain was very rocky, as can be seen with this large track tractor. Notice the suicide door on the pickup. (Courtesy of the Blue Ridge Parkway Archives.)

OCTOBER 1938 SLOPE GRADING. Section 1-S, near Tuggle Gap milepost 167, was under contract in March 1936 and was completed in December 1939. Here men are smoothing out the slope before the roller gets to their section. These men were probably with the CCC. They have on army uniforms and hard hats. These are some of the first type of hard hats developed. (Courtesy of the Blue Ridge Parkway Archives.)

CONCRETE MIXING SETUP IN 1948. Concrete was essential to the construction of culverts and bridges. This crew is building a culvert at the Peaks of Otter in Virginia. Johnson's Farm is a popular attraction here and was restored for visitors. (Courtesy of the Blue Ridge Parkway Archives.)

MAY 1937 CUTTING ROAD. Rocky Knob offers a wide variety of views, including flat tops and meadows. Here we see where the road was being cut through a very flat area. This project included 9.85 miles between Tuggle Gap and Virginia Highway 799. (Courtesy of the Blue Ridge Parkway Archives.)

SHE'LL BE COMING 'ROUND THE MOUNTAIN. The rolling valleys of Virginia are seen from this pretty view on the parkway yet to be paved sometime in the late 1930s. Once the roadbed was graded and leveled, workers had to build up the bed with crushed gravel. Then asphalt was poured over the roadbed. (Courtesy of the Blue Ridge Parkway Archives.)

CCC Shelter Overlook in 1938. The Rocky Knob Park shelter overlook was built by the CCC in May 1938. This shelter provides a safe spot from ever-changing weather to the hikers at Rocky Knob. The CCC would have built picnic shelters, park benches, and fences and done most of the landscaping for Rocky Knob. (Courtesy of the Blue Ridge Parkway Archives.)

Youth Conservation Corps in 1973. Youth Conservation Corps employees help replace rotting fences, mow, trim shrubs, and do general landscape maintenance under the supervision of the National Park Service (NPS). This community service is valued by the NPS and by the public. It provides essential training to young adults and exposes them to public careers. (Courtesy of the Blue Ridge Parkway Archives.)

ROCK QUARRY IN ROANOKE, VIRGINIA. Located near Roanoke, Virginia, this rock quarry would have been set up early in the process to crush rock for the roadbed being constructed. An outside contractor would have hauled this equipment to the site and set it up to operate. (Courtesy of the Blue Ridge Parkway Archives.)

TUNNEL EXCAVATION IN 1938. Tunnel excavation began in June 1938 for one of the many tunnels along the parkway. The south portal of this tunnel at milepost 369.3 below Craggy Gardens in North Carolina was very rocky. Hand drilling maintained more control over the excavation than with dynamite, which could have destabilized the area. (Courtesy of the Blue Ridge Parkway Archives.)

UNSEALED TUNNELS IN JANUARY 1939. The Twin Tunnels leaked water and developed icicles before they were sealed. The icicles were very large, indicating that the cold spell had lasted for at least several days. Near Buck Creek Gap at milepost 344.5, this section was completed sometime in 1939. Rock facing was placed on the outside of each tunnel. (Courtesy of the Blue Ridge Parkway Archives.)

Looking From Tunnel in June 1938. The blasted rocky mountain side is seen here from the portal of the tunnel at milepost 349 in North Carolina. This is called the Rough Ridge Tunnel and has a height of 21 feet and 6 inches. Green Knob overlook is one mile south of this tunnel with a view of the Catawba River Valley. (Courtesy of the Blue Ridge Parkway Archives.)

COMPLETED TUNNEL IN THE 1940s. Once the tunnels were completed, only one tunnel existed in Virginia. The others were located in North Carolina. Although the maximum speed limit on the parkway is 45 miles per hour, the average driver slows down to admire the engineering feat of tunneling through the mountains. (Courtesy of the Blue Ridge Parkway Archives.)

FANCY GAP MAINTENANCE FACILITY IN 1940. Eleven major maintenance areas were constructed, including this one at the Fancy Gap interchange. Each facility contained the field headquarters of the foreman and district ranger, as well as garage facilities for the cars, trucks, and heavy equipment. An FM radio transmitter was located at each district headquarters to keep in touch with the superintendent's office, rangers on patrol, and maintenance crews. (Courtesy of the Blue Ridge Parkway Archives.)

ACCESS RELOCATION IN JULY 1936. Project 2M was in process on July 25, 1936, and required the relocation of this access ramp, located near milepost 340. Earth movers had to chip away at the bank and remove several tons of soil to develop this access ramp to the standard specifications. (Courtesy of the Blue Ridge Parkway Archives.)

VIEW OF NORTH CAROLINA 26 INTERSECTION IN AUGUST 1935. Seen in this photograph on August 15, 1935, is the intersection of the Blue Ridge Parkway and North Carolina State Highway 26. This section, 2L, was constructed between July 18, 1937, and October 30, 1939, from McKinney Gap to Gooch Gap. The distance was 8.83 miles. (Courtesy of the Blue Ridge Parkway Archives.)

RAKES MILL POND IN JULY 1936. Parkway construction near Rakes Mill Pond was completed in November 1938. Section 1R was between Virginia State Road 793 and Tuggle Gap for 10 miles of the Blue Ridge Parkway. Great summer weather shows the work crews conducting grading, hauling, and hand work at this location. Large dirt haulers were necessary to carry in extra soil and gravel. (Courtesy of the Blue Ridge Parkway Archives.)

EARLY VIEW SHED CONSTRUCTION IN 1947. View shed construction in 1947 was rather difficult because of the rugged mountain terrain and higher elevations seen in this North Carolina view. This photograph was taken in February 1937, just after the chestnut blight, which left chestnut trees all over the United States looking like ghost trees. (Courtesy of the Blue Ridge Parkway Archives.)

CONSTRUCTION IN AUGUST 1938. A 100-foot cut through a bank is shown in this photograph at milepost 319.6 between Highway 221 and McKinney Gap in North Carolina. Construction began on March 3, 1938, and was completed in October 1941. A large machined shovel is doing the work, and a pickup truck follows, ready to haul some of the debris away. (Courtesy of the Blue Ridge Parkway Archives.)

WORM FENCE NEAR PUCKETT CABIN. Dead chestnut trees were recycled into worm fencing all along the Blue Ridge Parkway by the CCC. Crews would saw down the trees and split them into usable lengths. The Ground Hog Mountain picnic area contains about six examples of fence types. A lookout tower used to watch for wildfires still stands. (Courtesy of the Blue Ridge Parkway Archives.)

YOUTH CONSERVATION CORPS. The Youth Conservation Corps (YCC) is the most well-recognized youth program in the National Parks System. It has been instrumental in introducing young Americans to conservation opportunities in national parks since the program was created in 1970. This program was created through a partnership between the U.S. Department of the Interior and U.S. Department of Agriculture Forest Service. Established on August 13, 1971, through Public Law 91-378, the YCC was a three-year pilot program, with the intention of achieving several objectives. The most important objective was to take young adults from different social, economic, racial, cultural, and gender backgrounds and place them in an environment where they could cultivate work, social, and educational skills. The YCC became a permanent organization in 1974. (Courtesy of the Blue Ridge Parkway Archives.)

Three

HEAVY EQUIPMENT

Horse-drawn equipment made up the early machines of road building until the introduction of steam power. By the time the Blue Ridge Parkway was being built in the 1930s, gasoline-, electric-, and diesel-powered machines were in use.

This chapter will highlight some of the equipment used to build the Blue Ridge Parkway.

Benjamin Holt of Stockton, California, perfected the first successful crawler tractor in 1904. The Caterpillar trade name was first applied to such a machine the following year. Holt's company photographer thought of a giant caterpillar when he saw the machine moving on its continuous steel crawlers equipped with cleats, or shoes, for traction.

In 1925, the Caterpillar Tractor Company was formed through a merger of Holt's firm and bitter competitor C. L. Best Tractor Company of San Leandro, California. Caterpillar eventually consolidated its operations at the former Holt plant at Peoria, Illinois, and Peoria has been Caterpillar's world headquarters ever since. The Caterpillar Thirty, so designated because of its 30-horsepower engine, was introduced by Best in 1921.

The term *dozer* or *bulldozer* is often applied to a crawler tractor. More accurately, dozer refers to the blade on the front of the tractor that is used to gather, push, and spread material. Faster self-propelled, rubber-tired machinery rendered them largely obsolete for this work starting in the 1940s and 1950s.

By the 1960s, self-propelled motor scrapers rendered the much slower tractor-drawn scrapers obsolete for nearly all construction applications because of their speed and maneuverability. Many continued to be used for their original purpose of agricultural land leveling, as well as occasional use in construction work where severe conditions made motor scrapers impractical.

The earliest known dump trucks date to 1904. Early dump truck capacities were only a couple of cubic yards, and today tandem-axle trucks like these carry 10 to 12 cubic yards.

SHOVEL DOZER AT BORROW PIT. A bulldozer is a very powerful crawler (caterpillar-tracked tractor) equipped with a large blade. The first dozers were equipped from farm Holt tractors that were used to plough fields. The blade peels layers of soil and pushes it forward as the tractor advances. Notice this one does not have a cabin to shield the driver from the elements. The borrow pit literally contains excavated material that has been taken from or borrowed from one area to be used as fill at another location. At each stage of the construction where fill dirt was needed, these borrow pits were established to help stage the construction. Projects the size of the Blue Ridge Parkway have to be done in sections. It took over 52 years to complete 45 sections of the Blue Ridge Parkway. (Courtesy of the Blue Ridge Parkway Archives.)

MOWING EQUIPMENT IN 1973. Tractors with bush hogs and grass blades are essential to the upkeep and grooming of the Blue Ridge Parkway. During the warm months of the year, mowing is a constant job for the maintenance crews. This crew is getting ready to mow at Gillespie Gap in North Carolina. (Courtesy of the Blue Ridge Parkway Archives.)

SHOVEL AND TRUCK IN THE LATE 1930S. Shovels became a mainstay of road construction. The cab sits on a swivel so that the driver can turn 360 degrees to access rock that has to be hauled away once blasted. Heavy-duty army trucks were used to haul the large loads to other locations or to crushers for recycling. (Courtesy of the Blue Ridge Parkway Archives.)

QUALITY CONTROL CREWS IN 1973. Asphalt repair requires a lot of preparation. Before the process begins, the determination has to be made as to the type of materials that will be used and as to the degree of thickness the materials will be covering. Surveys have to be conducted to determine the width of the road. (Courtesy of the Blue Ridge Parkway Archives.)

PARKWAY REPAIR IN 1978. Repairing the parkway near Altapass, North Carolina, in 1978 required a large dump truck with a load of gravel. Gravel provides a firm foundation for the buildup of the roadbed. Once it is packed and leveled, asphalt will be added. (Courtesy of the Blue Ridge Parkway Archives.)

DUMP TRUCK WITH GRAVEL SPREADER. In October 1976, a section between Little Switzerland and Gillespie Gap of the Blue Ridge Parkway had to be paved. This dump truck is pulling a gravel spreader with an operator standing over the spreader to move the levers, allowing gravel to spill smoothly. (Courtesy of the Blue Ridge Parkway Archives.)

THREE-WHEEL ROAD ROLLER. Once gravel is spread, it must be compacted to a certain thickness and smoothness. Here an operator sits atop a compactor without a hood for shelter from the elements. The first rollers were horse drawn on dirt roads. The wheels were pneumatic. (Courtesy of the Blue Ridge Parkway Archives.)

SEAL COATING. Asphalt sealer has to be sprayed over the layer of asphalt to seal any cracks from the elements. Here is a one-half-ton truck with a large tank of sealer spraying a section of the Blue Ridge Parkway between Little Switzerland and Gillespie Gap in North Carolina in October 1976. (Courtesy of the Blue Ridge Parkway Archives.)

TRACK TRACTOR HAULING IN 1936. This project near Doughton Park in Alleghany County, North Carolina, covered rocky ground. Dynamite had to be used to blast the rock to pieces small enough to be picked up by the shovel in the foreground and loaded onto this track tractor. Track tractors grew bigger and more powerful over the years; they also acquired raised cabins to protect the driver. Experienced miners lived in the mountain regions of Virginia and North Carolina. It's more than likely that some of the explosive handlers were experienced in the mining industry. By the late 1900s, nitroglycerine had replaced basic gunpowder for explosions. During the early 20th century, the first experiments were done with charges. In 1930, the U.S. Bureau of Mines was formed to upgrade safety. In 1935, DuPont introduced the first commercially successful non-nitroglycerin ammonium nitrate (AN) blasting agent. Thus during the early construction period of the Blue Ridge Parkway, blasting technology was expanding. This greatly increased the accuracy of carving the roadbed out of the mountainsides. (Courtesy of the Blue Ridge Parkway Archives.)

Paving Traffic Jam in 1976. During peak season, traffic jams develop easily, especially when road construction is in process. Although inconvenient, paving has to be completed when the weather is just right. When temperatures drop below 50 degrees, asphalt will not cure properly. Rationing of gasoline during this time period interfered with the routine maintenance of the parkway. (Courtesy of the Blue Ridge Parkway Archives.)

ROAD-STRIPING EQUIPMENT. This three-man crew operates a yellow striping truck in June 1973. A double yellow line is placed down the center of the parkway. Earlier years of the parkway indicate that only a single white line was placed in the center of the road. As traffic increased and road laws changed, a double yellow line was adopted. (Courtesy of the Blue Ridge Parkway Archives.)

MOVING A BRIDGE GIRDER. This prestressed concrete bridge girder took a lot of maneuvering along the parkway in June 1959. The construction crew had to handle this monster around curves and over hills to get it in place. A crane can be seen over the treetops waiting to position the girder over cement pillars. (Courtesy of the Blue Ridge Parkway Archives.)

CONSTRUCTION TRAFFIC JAM. Asphalt paving and mowing seem to be the holdup in this traffic jam in 1976. Americans have always flocked to the parkway in the fall to see the changing leaves. A glitch in the timing of paving caused traffic to build up at this intersection, even though this work was probably completed during the typical work week. (Courtesy of the Blue Ridge Parkway Archives.)

BUCKET EXCAVATOR IN LATE 1930S. Large excavators used by contractors helped to speed along the paving process of the Blue Ridge Parkway. This excavator is loading large boulders into a track tractor with a dumpbed near the Doughton Park area of the parkway. Cabins sit on top of a swivel base with tracks that grip the ground and move the vehicle in all directions. (Courtesy of the Blue Ridge Parkway Archives.)

Shovel Excavator Loading Truck. Diesel-powered excavators replaced steam shovels in the 1930s. This swivel shovel is loading large boulders into a late-1930s-model pickup with dual wheels. The Department of the Interior was sponsoring this section, as can be seen by their logo on the shovel. (Courtesy of the North Carolina Archives.)

W. H. ARMENTROUT OF HARRISONBURG, VIRGINIA. W. H. Armentrout won several contracts for road construction on the Blue Ridge Parkway. Photographed here is a section in Virginia. Armentrout provided manpower, equipment, and asphalt paving. After their job was completed, the CCC would landscape the sides of the roads and plant native shrubs. (Courtesy of the North Carolina Archives.)

ALBERT BROTHERS CONTRACTORS, INC. Albert Brothers of Salem, Virginia, were another contributor to the construction of the Blue Ridge Parkway. Their contracts covered sections in Virginia and North Carolina. All of the equipment in this photograph is outfitted with tracks, which are best for gripping the earth in difficult terrain. (Courtesy of the North Carolina Archives.)

MOVING INTO THE CHESTNUT GRAVEYARD. Difficult does not even begin to describe the conditions that contractors had to move through in the early construction of the Blue Ridge Parkway. Ruts several feet deep can be seen in this wide-angle photograph taken on this section in Virginia. Large dump trucks pulling loads of soil, with track tractors following them, were moving through a section of blighted chestnut trees. (Courtesy of the North Carolina Archives.)

SLINGING MUD IN VIRGINIA, c. 1936.
Trucks used to be light enough to push.
Here we see several men pushing a pickup
out of ruts that are at least 2 feet deep.
The person behind the wheel could count
on a good mud bath. (Courtesy of the
North Carolina Archives.)

ASPHALT PAVING IN VIRGINIA. This
small asphalt-paving operation must have
been demonstrating patching potholes.
Asphalt is fed into the paving unit and
then spread on the surface. Newer models
are fed by a large dump truck and span the
width of the road. (Courtesy of the North
Carolina Archives.)

BUCKET OF TROUBLE, c. 1936. Curiosity snagged these two ladies in the early days of construction in Virginia. They are sitting in the bucket of a diesel shovel. Visitors could not wait for the road to be completed, and many would take a walk along the unfinished roadbed to get a glimpse of what was to come. Although the country was in a Depression, fashion and style did not stop. Women's fashions moved away from the brash, daring style of the Roaring Twenties toward a more romantic, feminine silhouette. The waist was restored to its proper position, hemlines were dropped, there was renewed appreciation of the bust, and backless evening gowns and soft, slim-fitting day dresses became popular. (Courtesy of the North Carolina Archives.)

A 1930s Dump Truck. Hauling tons of soil and rock, dump trucks were invaluable during the construction of the parkway. The early ones did not have a cab to cover the driver. The bed was equipped with a hydraulic lift for pouring loose gravel and dirt. One axle was under the cab, and one or more could be mounted under the bed. (Courtesy of the North Carolina Archives.)

Caterpillar Tractor. Organized on April 15, 1925, Caterpillar equipment is by far the most well-known construction equipment in the United States, and the company's early equipment ranged from diesel and gas turbines, to natural gas. The Caterpillar tractor inspired the first tank used in World War I. Caterpillar advertised that one of their tractors replaced eight bullock teams—112 animals. (Courtesy of the North Carolina Archives.)

RUGGED DUMP TRUCK. A tractor engine and a dump bed connected by a frame and rolling on axles describes this simple piece of equipment. This driver was in danger of flying objects when driving this truck. It was a workhorse and could haul loads of high tonnage. (Courtesy of the North Carolina Archives.)

EARLY-1940S CHEVROLET TRUCK. Cofounded by Louis Chevrolet and William C. Durant in 1911, Chevrolet is the most popular brand of General Motors vehicles sold in America. The "bow tie" logo first used in 1913 by Louis Chevrolet has become the standard symbol for Chevrolet. Hauling supplies probably was the purpose of this vehicle. (Courtesy of the North Carolina Archives.)

PRE-1930s PNEUMATIC ENGINE. Stored energy in the form of compressed air, nitrogen, or natural gas enters the sealed motor chamber and exerts pressure against the vanes of a rotor. Much like a windmill, this causes the rotor to turn at a high rate of speed. Reduction gears are used to create high torque levels. This would probably have been used with a jackhammer. (Courtesy of the North Carolina Archives.)

DIESEL 50 TRACTOR. After 1933, all tractors were offered with rubber tires. In 1940, ninety-five percent were sold with rubber tires. Before 1931, tractors had steel wheels with lugs jutting out. This tore up the ground like a plugger does today. Caterpillar tractors were advertised to do five times as much work as a horse. (Courtesy of the North Carolina Archives.)

Four

ROCK SLIDES AND WEATHER

For over a billion years, the Blue Ridge Mountains have been affected by environmental factors. Two landmasses collided more than 1.1 billion years ago, forming the ridges we see today. Though they were once similar to the Colorado Rockies, erosion has weathered down the peaks to today's heights.

Summer thunderstorms bring torrents of rain to the Blue Ridge Mountains. In the winter, freezing and thawing water in crevices brings occasional rock slides that highlight the erosion processes in these mountains. Occasional catastrophic events like floods, hurricane-force winds, blizzards, and ice storms can change the face of the mountains overnight.

Severe weather has affected the parkway from the beginning. In 1940, the Georgia–South Carolina Hurricane caused flooding and landslides. You will see several photographs of the devastation that this storm created. This storm was just a little smaller than Hurricane Hugo, which hit in 1989.

In 2004, Hurricane Frances washed out several sections of the Blue Ridge Parkway. The remnants of Hurricane Frances crossed the region on September 7–9, causing multiple rock slides and completely washing out some portions of the parkway in Western North Carolina. Estimates of damage to the road and visitor facilities are over $11 million. Major rock slides and the most severe damage occurred between mileposts 322 and 349, and some of this portion of the parkway was closed for up to a year. Extensive power outages closed several of the visitor centers.

Elevation varies on the parkway from 650 feet at the James River in Virginia to over 6,050 feet south of Mount Pisgah. The weather changes quickly because of these altitude differences, sometimes without warning.

"An open Parkway is a safe Parkway" is the motto of the National Park Service. Safety is of the highest concern of park officials. If the road is closed, it is not safe. Continuous environmental assessments are made to ensure the safety of the visitors and to maintain the road.

SNOW DAMAGE IN 1974. Very beautiful, heavy snows create havoc on the parkway. This December 11, 1974, storm broke limbs and trees all over the North Carolina section of the Blue Ridge Parkway between milepost 320 and milepost 330. Once damage has happened, maintenance crews work until opening in the spring to make the road safe again. (Courtesy of the Blue Ridge Parkway Archives.)

ROCK SLIDE AT CRAGGY GARDENS. Rock slides occur throughout the mountain roads, including the Blue Ridge Parkway. They usually occur after heavy rains and in areas that are prone to slides. Layered slate rock frequently becomes unstable, causing slides. Drivers should be cautious when entering these zones. (Courtesy of the Blue Ridge Parkway Archives.)

WASHOUT NEAR ROCKY KNOB IN 1971. On October 24, 1971, the road washed out near Rocky Knob milepost 131.8. Unstable banks near roadbeds often wash out after heavy rains. This section of the road had to be closed until repairs could be made. Rainfall in the Blue Ridge Mountains is known to be heavy and sometimes causes major damage. (Courtesy of the Blue Ridge Parkway Archives.)

STREAM DAMAGE IN 1940. Below Grandfather Mountain, stream damage occurred from heavy rains. This section of the parkway lies between Gooch Gap and Big Laurel Gap. Completed in July 1940, it is still a very winding road, passing through the wilderness in the Blue Ridge Mountains. (Courtesy of the Blue Ridge Parkway Archives.)

CALDWELL COUNTY ROCK SLIDE. In August 1940, heavy rains caused damage to the Blue Ridge Parkway in the form of rock slides and fallen trees. Large boulders came crashing down the side of the mountain near 221 in section 2H, which had been completed in July of the same year. It was hard to keep the road open with all of the extra maintenance required to clear the debris. (Courtesy of the Blue Ridge Parkway Archives.)

FILL WASHOUT IN 1940. Fill dirt used to build the roadbed washed out near milepost 336 in North Carolina. Sometimes fill dirt is not as settled as the local terrain. Hurricane Georgia–South Carolina was devastating to mountain counties in Tennessee and North Carolina. It made landfall on August 11 and its effects lingered for five or six days. (Courtesy of the Blue Ridge Parkway Archives.)

LANDSLIDE AT MILEPOST 320 IN 1940. The Hurricane Georgia–South Carolina in August 1940 caused heavy rains that sparked disastrous flash floods. The floods inundated much of Tennessee, the Carolinas, and northern Georgia. Press reports stated that 30 more people died in the floods, and it caused millions of dollars in damage. Over 12 inches fell in the mountains of North Carolina. (Courtesy of the North Carolina Archives.)

ROCK SLIDE IN 1957. This rock slide near milepost 421 in North Carolina happened in April 1957. Half of the mountain came down in the middle of the road and stopped all traffic until crews could clear the road. Notice the car in the distance compared with the size of the boulders. Other factors contributing to rock slides are erosion, earthquakes, weak slopes, snow and rainfall, and even vibrations made by equipment like passing cars. This is probably a "shallow landslide" caused by slope failure on the side slope of the parkway. Winter weather in March 1957 was unseasonably cold and probably had more frozen precipitation. When the spring thaw came in April, the mountains saw more runoff than normal, which caused landslides. (Courtesy of the Blue Ridge Parkway Archives.)

Five

WATER AND ARCHITECTURAL FEATURES

Distinctive architectural water structures of the Blue Ridge Parkway include the collection of bridges and grade-separation structures. They allow the parkway to cross streams and other roads. Many of these 170 structures are arch-type bridges with rustic stone facades, allowing them to blend into the mountain landscape. Much of the stone was gathered locally by the contractors. Some of the bridges are steel and concrete.

During the early design phase of the parkway, the Bureau of Public Roads (now the Federal Highway Association) collaborated with the National Park Service on the design of the structures.

Stone facing for the bridges and grade-separation structures became the hallmark style of architecture used by the National Park Service. Rock was either acquired from local quarries or obtained from the rock cuts created during the road construction. The stone varies on the 469-mile route depending on the geology of the surrounding area.

An arch bridge is a bridge with abutments at each end shaped as a curved arch. Arch bridges work by transferring the weight of the bridge and its loads partially into a horizontal thrust restrained by the abutments at either side.

Several different arch-type bridges were built along the parkway. The landscape dictated the length and span of the arches. An elliptical arch was used when there was sufficient horizontal clearance. This allowed the arch to be carried all the way to the ground. In some situations, circular arches were employed for narrower spans. Crossing streams and rivers sometimes called for a skewed curved bridge on the diagonal.

Culverts and drains were formed in similar fashion, with stones covering the fronts and sides of the culverts. The stonework was not merely cosmetic but helped form the concrete frame.

The parkway has 26 tunnels. They are arched shaped and stone faced like the bridges.

Several dams were built along the parkway for holding ponds, fishing ponds, and reservoirs.

THE FIRST BOX CULVERT. This box culvert measures 4 feet by 6 feet and is located at milepost 229 in North Carolina just south of Cumberland Knob. More than likely this is one of the first culverts built on the Blue Ridge Parkway. A contractor would have built this culvert, not the CCC. Box culverts were designed to be sturdier than corrugated piping and would hold a higher volume of water. Most of the culverts found in the early construction of the parkway are natural-bottom culverts with sand, gravel, and dirt of the native creek beds as the bottoms. They were three-sided structures. An environmental plus to these culverts is that they allow fish passage. For the trout streams found in the Blue Ridge Mountains, this is important. (Courtesy of the Blue Ridge Parkway Archives.)

LINVILLE OVERPASS, c. 1939. Located near milepost 317.5, this overpass was probably completed sometime in 1940. Notice the old tractor crane in the background. All bridges and overpasses had to be covered with native stone. A cement factory was set up on-site. (Courtesy of the Blue Ridge Parkway Archives.)

HEADWALL CULVERT IN 1936. Built near the Virginia and North Carolina line between 1935 and 1936, this is a great example of a headwall culvert. It is covered with native stone and has an arch over the single opening. Drainage is the main purpose of this culvert. (Courtesy of the Blue Ridge Parkway Archives.)

DOUBLE-BOX MODEL. Engineering models of the culverts were built by the engineers. This model is of a double-box stone masonry culvert and was built in 1935. Native stone was laid on the face of each culvert to give it a more uniform, finished look. (Courtesy of the Blue Ridge Parkway Archives.)

MODIFIED TYPE 3 CULVERT. Grading complete, this culvert was modified and displays side wings of stone walls. Located in Virginia at milepost 148, between Pine Spur overlook and Smart View, this culvert was completed in December 1936. A small stream flows through this culvert, which explains why extra space was needed. Culverts are engineered with many factors in mind. These factors are normal stream flow and heavy flow after rainfall. Sometimes the size of the stream and fish species has to be considered to allow enough space for the fish migration to spawn. Formerly, these streams would have been called a "ford," with either a low water bridge or nothing at all to drive over. When flooding occurs, water goes over the low water bridge. The idea behind culverts and low-water bridges is that they are safe under normal conditions to cross. Under flooding conditions, the pathway may be impassable. (Courtesy of the Blue Ridge Parkway Archives.)

THE FISHING HOLE. Snapped in September 1953, this photograph shows a beautiful fishing hole on the parkway. Some water projects were natural to the parkway and were only landscaped, while damming up streams created others. This one has a wonderful picnic area and parking lot. The designers of the parkway wanted to create "atmosphere," and they did this quite often with water features like this small pond. Visitors would have to have a permit to fish here. Some of the ponds and streams are stocked. Special fishing spots in Virginia include Mill Creek, Abbott Lake, Little Stoney Creek, Otter Lake and Creek, Rock Castle Creek, Little Rock Castle Creek, and Chestnut Creek. North Carolina fishing spots include Trout Lake, Upper Boone Fork (upstream from Price Lake), Lower Boone Fork, Cold Prong Branch, Laurel Creek, Sims Pond and Creek, and Camp Creek. Fishing is not permitted from the dam at Price Lake, from the footbridge in the Price Park picnic area, or from the James River Bridge. (Courtesy of the Blue Ridge Parkway Archives.)

DOUBLE-BOX CULVERT IN 1936. A rather heavy-flowing stream flows through this double-box culvert near milepost 150 in Virginia. Between Pine Spur and Smart View, this double culvert was completed sometime in 1936. The old Ford Model T truck has spoke wheels and was based on the car Model T. They were mass-produced between 1925 and 1927. (Courtesy of the Blue Ridge Parkway Archives.)

DOUGHTON PARK WATER TOWER. The storage tank at Doughton Park was erected in 1938 and provides storage today. This is a great lookout point for photographers. Doughton Park used to be called the Bluffs but was renamed after Sen. Robert Doughton, who was instrumental in the parkway path through Alleghany County, North Carolina. (Courtesy of the Blue Ridge Parkway Archives.)

EARLY NORTH CAROLINA DAM. The Tennessee Valley Authority (TVA) was another public works program and built dams in Eastern Tennessee and Western North Carolina. This dam is probably the Lake Junaluska Dam, c. 1930. The parkway has 14 dams; the parkway dams are not as large as the electrical dams built by the TVA but are just as impressive. (Courtesy of a private collection.)

DAM CONSTRUCTION, C. 1930. Located in the Western North Carolina mountains, this dam was constructed sometime in the 1930s during the boom in public works construction. They provided power for the electricity that would be supplied to homes and businesses, and they provided a good supply of water. (Courtesy of a private collection.)

PRICE PARK DAM LOWER SIDE. The Julian Price Memorial Park was named after the late Julian Price, who was for many years president of the Jefferson Standard Life Insurance Company of Greensboro, North Carolina. The park is composed of 4,200 acres. The company and Price's son and daughter, Ralph C. Price and Mrs. Joseph McKinley Bryan, cooperated in the acquisition and dedication of this property as a public recreation area. It is one of the largest developed campground areas on the parkway with 129 tent and 68 RV sites, and those on Loop A are located next to Price Lake. Interpretive programs, fishing, boat rentals, and an extensive trail system, including the Tanawha Trail across the face of Grandfather Mountain, make for a delightful parkway visit. This section of the parkway is located next to the Moses Cone Memorial Park. Blowing Rock and Boone are the nearest towns. (Courtesy of the Blue Ridge Parkway Archives.)

PRICE PARK DAM. Located at milepost 296.8 on the Blue Ridge Parkway is the Julian Price Lake Dam. Constructed in 1960, this photograph shows the dam from the upper side. Today it is one of the most popular spots on the parkway for camping and fishing. (Courtesy of the Blue Ridge Parkway Archives.)

SIMS POND AT MILEPOST 295.9. Sims Pond and Creek are located next to the Price Park and Lake. Parking and accessible trails lead the visitor to this beautiful pond. A one-mile trail has been developed around the pond. Fishing is allowed. (Courtesy of the Blue Ridge Parkway Archives.)

LINVILLE RIVER BRIDGE. Built in 1941 at milepost 316.6, the Linville River Bridge is the largest stone structure on the Blue Ridge Parkway. Distinctive of parkway construction are the 168 structures that allow visitors to pass over streams, rivers, and ravines; they are covered with rustic stone to blend in with the landscape. Others are sleek modern steel-and-reinforced-concrete structures. The old stone arch bridges were constructed by erecting stone arch rings, abutments, and abutments and spandrel walls, and then pouring concrete on a large network of steel reinforcing rods. Not only decorative, the stonework provides form for the concrete frame. All but one of the overpasses are stone-faced-arch structures. Stone-facing structures became a hallmark of the National Park Service during construction. Generally, stone was obtained from local quarries, creating the variance of stone types depending on the location of the structure. Different shapes were employed, with choice being dictated by the length and span of the ravine or river. (Courtesy of the Blue Ridge Parkway Archives.)

LOOKING UP! LOOKING DOWN! The Linville River Bridge is one of the most awe-inspiring engineering structures on the Blue Ridge Parkway. Here, in June 1940, visitors are touring the new structure and wondering, "Where is the water?" Today a one-tenth-mile trail leads to the structure where the Linville River flows out of the Linville Gorge. (Courtesy of the Blue Ridge Parkway Archives.)

ELLIPTICAL ARCH BRIDGE. A winter blanket covers this skewed elliptical arch bridge that has a stone face. The single arch provides water flow for a small creek. Some of the arches were skewed on a diagonal so as not to disturb the flow of the roadway. This created an amazing bridge. (Courtesy of the Blue Ridge Parkway Archives.)

LINVILLE FALLS. Surrounded by towering hemlocks, the Linville Falls were created by a rare geological phenomenon in which older rock overlay younger rock—evidence of ancient collisions between the North American and African continents. This is the lower basin of the falls. Purchased with funding from John D. Rockefeller, Linville Falls is a favorite spot for the author. (Courtesy of the Blue Ridge Parkway Archives.)

Erwin's View of Linville Falls. Milepost 316.4 boasts the powerful Linville Falls, which, as Representative Doughton said during his negotiating for the path of the parkway, was "Chiseled by the Omnipotent Architect of the World . . . [God] has carved the most outstanding display of nature known to all creation." Now one of the largest natural attractions of the parkway, Linville Falls can be viewed from several balconies. Instead of several falls on one trail, this site offers six different views and three different trails for one superb waterfall. Linville Falls has a double cascade with a vanishing act between the two falls. Headwaters of the Linville River are at Grandfather Mountain. It flows to the Catawba Valley and into the Catawba River. The Cherokees called the area "Eeseeoh," which means "river of cliffs." Settlers called the river and the falls "Linville" to honor the explorer William Linville, who in 1766 was attacked and killed in the gorge by Native Americans. (Courtesy of the Blue Ridge Parkway Archives.)

Six

INTERPRETIVE SITES

Work began on the Blue Ridge Parkway on September 11, 1935, near Cumberland Knob in Alleghany County, North Carolina. Now the parkway is the most visited park in the United States and runs for 469 miles. Construction of the parkway took over 52 years, and it was completed in 1987.

The parkway is marked with milepost interpretive markers beginning with 0 at the terminus of Skyline Drive in Virginia and ending at milepost 469 at Oconaluftee Village in Cherokee, North Carolina. Interpretive sites were saved throughout the 469 miles to showcase the Appalachian heritage and crafts.

There are many surviving examples of early European pioneer structures along the parkway. Starting at milepost 5.8 at the Humpback Rocks Visitor Center and Mountain Farm exhibit, the easy Mountain Farm Self-Guiding Trail takes one through a collection of 19th-century farm buildings where—in the summer months—there are often living history demonstrations.

Other interpretive cabins include the Trail Cabin (milepost 154.6), Puckett Cabin (milepost 189.9), Brinegar Cabin (milepost 238.5), Caudill Cabin (milepost 241), and Sheets Cabin (milepost 252.4), all of which are 19th-century log cabins illustrating the isolated existence of mountain residents. These cabins show the visitor the efforts of the original park planners to save log structures as opposed to other types of larger farmhouses they found. The Jesse Brown Farmstead (milepost 272.5) consists of a cabin, springhouse, and the relocated Cool Springs Baptist Church.

Brinegar Cabin, near milepost 238, is the only building on the National Register of Historic Places. This cabin became an interpretive site for demonstrating early American textile skills, such as weaving and dyeing.

There are 250 overlooks on the parkway. Each one has a sign stating the name of the overlook, elevation, and sometimes a description of the scene.

The most recent addition to the parkway is the Blue Ridge Music Center, which was constructed between the Fancy Gap milepost and the Galax milepost. Photographs and exhibits of early bluegrass bands in the area are highlighted in the visitor center. An amphitheater hosts different events throughout the tourist season.

BRINEGAR CABIN, C. 1941. Before restoration, the Brinegar Cabin was in dire need of repair. Located at milepost 238.5, the cabin once belonged to Martin and Catherine Brinegar. They were self-sustained on the mountaintop with only a couple of crops that they sold at market each year. Their descendants meet each August and hold a community festival at the cabin. (Courtesy of the Blue Ridge Parkway Archives.)

JAMES MARTIN AND CAROLINE JOINES BRINEGAR. Martin Brinegar was a descendant of the first Brinegar to immigrate to America from Europe. Martin and Caroline were married February 9, 1878, in Alleghany County, North Carolina. Their children were Susy Alice, Sarah Lurene, John William, and Mac Henderson Brinegar. (Courtesy of the Blue Ridge Parkway Archives.)

BRINEGAR CABIN AFTER RESTORATION. Now listed on the National Register of Historic Places, Brinegar Cabin was built around 1885 and restored and opened to the public in 1957. Several outbuildings were also included. This photograph was taken in August 1955 and shows the attention to detail given to the restoration. (Courtesy of the Blue Ridge Parkway Archives.)

SORGHUM PRESS DEMONSTRATION. Sweet sorghum is a syrup made from the juice of sorghum cane. In years past, it was an important source of sweetener for breads, desserts, and drinks. During the 1850s, it came into prominence in the United States. By 1888, total U.S. production was 20 million gallons. (Courtesy of the Blue Ridge Parkway Archives.)

APPLE BUTTER DEMONSTRATION. This demonstration of apple butter production took place in October 1961 at Mabry Mill. Many days before making this spread, several gallons of apples have to be picked, peeled, and cooked to applesauce consistency. Once in the kettle, the solution of applesauce and sugar are cooked for about eight hours. Then the cinnamon flavoring is added before being transferred to jars. It was a popular way of using apples in Colonial America well into the 19th century. The term "butter" refers to the thick, soft consistency and its use as a spread for breads. Typically seasoned with cinnamon, cloves, and other spices, apple butter may be used as a side dish, an ingredient in baked goods, or as a condiment. The Pennsylvania Dutch often include it as part of their traditional seven sweets and seven sours dinner table array. In the Blue Ridge Mountains, it is a family event to make apple butter. Every family member can take part in the process. (Courtesy of the Blue Ridge Parkway Archives.)

HOG KILLING IN 1953. Hogs provided meat and fats to the mountaineer. Hogs were butchered and sliced into roast, hams, and sausage. The fat was extracted and mixed with lye to make soap. Weather played a big part, as it has to be cold enough to preserve the meat but not freeze the meat. (Courtesy of the Blue Ridge Parkway Archives.)

THE NORTH CAROLINA MINERALS MUSEUM. Located in Spruce Pine, North Carolina, at milepost 331, the North Carolina Minerals Museum was dedicated on June 17, 1955. This photograph of Junior McKinney was taken in May 1974. The museum provides an introduction to the importance of mining in the region. (Courtesy of the Blue Ridge Parkway Archives.)

GUIDED TOUR OF MABRY MILL. Park ranger Brenda Bowers used to give guided tours of Mabry Mill. Here a demonstration of fence building and supplies is shown to an enthusiastic crowd of visitors. In August 1975, the parkway was a very popular vacation destination. (Courtesy of the Blue Ridge Parkway Archives.)

MR. AND MRS. ROSCOE MATTHEWS. This photograph was published in the March 14, 1957, *Galax Gazette*. The Matthewses donated the family log cabin to the park for demonstrations. The Mabry house was destroyed before the mill was reopened and was replaced with the Matthews Cabin. (Courtesy of the Blue Ridge Parkway Archives.)

MR. AND MRS. ROSCOE MATTHEWS WITH SIGN. The log cabin donated by the Matthewses was built in 1869 by Samuel Monroe and Elizabeth Ann Matthews in the Mount Lebanon community of Carroll County, Virginia. The building was moved to the parkway in October 1956. (Courtesy of the Blue Ridge Parkway Archives.)

EARLY RECONSTRUCTION OF MABRY MILL. Reconstruction of Mabry Mill began in the 1940s. Here is a track shovel that is going to be used to dig out the pond in front of the wheel. The old white clapboard house of the Mabrys had already been destroyed, and the Matthews Cabin is being erected in the background. (Courtesy of the Blue Ridge Parkway Archives.)

Traffic Jam at Mabry Mill. Riders along the Blue Ridge Parkway looking at the fall colors in October 1961 caused a traffic jam at Mabry Mill. All of the parking lots were full. One can smell the country ham and apple butter just looking at this photograph. (Courtesy of the Blue Ridge Parkway Archives.)

WATER-WHEEL SHAFT. In June 1942, a 30-inch timber was cut for a water-wheel shaft. It was hand hewn by a craftsman working on the restoration project. The Mabry Mill, located at milepost 176.1, fell into disrepair after E. B. Mabry passed away. The mill went into operation in 1905 and suspended operation in 1935. (Courtesy of the Blue Ridge Parkway Archives.)

COMPLETED WATER WHEEL. In July 1942, this photograph was taken of the men who rebuilt the water wheel. From left to right are men identified only as Hylton, Goad, Scott, and McCarter. Because of their skills and craftsmanship, Mabry Mill is one of the most photographed mills in the world. (Courtesy of the Blue Ridge Parkway Archives.)

WINTER AT MABRY MILL. This photograph was taken in the cold of winter. Ice had frozen on the water wheel and flume. Before the mill opened to the public, the small lake had to be built. At the time of this photograph, the lake had not been developed. Mabry Mill, a historic gristmill, was opened to the public in the 1950s. (Courtesy of the Blue Ridge Parkway Archives.)

MABRY MILL POND. The National Park Service added the pond in front of Mabry Mill. What a great addition! Many great photographs have been taken with the pond. A rustic stone-covered bridge carries the road within feet of the mill and pond. The sides of the arch stand straight and are not curved. (Courtesy of the Blue Ridge Parkway Archives.)

MAXIMUM LIMIT BUSES. Checking stations were used in the early days of the parkway. Charter buses were only allowed with permits. This wonderful photograph was taken in July 1939 near milepost 360. Mount Mitchell is near milepost 360, which is a popular destination boasting the highest peak east of the Mississippi River. During the 1920s, buses began to replace streetcars because of the mechanical problems of the cable cars, which were also limited geographically. Thomas Buses of North Carolina is but one example; they produced their first school bus in 1936. The buses pictured here held between 30 and 40 people and were built sometime in the 1920s or 1930s. Notice these buses have three axles and probably were produced by a division of General Motors called Yellow Coach. The first 50-passenger bus wasn't developed in the United States until 1948. The parkway today is bus accessible, but large coach buses must have a special permit. (Courtesy of the Blue Ridge Parkway Archives.)

ORLEAN HAWKS PUCKETT OUTBUILDING. This is a real-photo postcard of what was likely Orlean Puckett's corncrib, which dates back to the 1800s. She and her husband, John Puckett, were married before the Civil War. "Aunt" Orlean took to midwifery in her mid-forties. She is well known for delivering over 1,000 babies during her lifetime. (Courtesy of the Blue Ridge Parkway Archives.)

RETAINED PUCKETT CABIN IN 1939. This cabin was retained by the Department of the Interior to tell the story of Orlean Puckett. Taken on July 25, 1939, this photograph shows that the cabin was in great condition with weeds growing up on the sides. Puckett was a midwife who ironically lost all 24 of her own children before they reached toddler age. (Courtesy of the Blue Ridge Parkway Archives.)

PUCKETT BUILDING GROUP IN 1939. This photograph shows other buildings that existed on the Puckett property. Orlean lived here until the parkway bought her property and construction began. She died one year later. Even in snow and ice, she never missed a birth. She hammered nails in her shoes to grip the ice. (Courtesy of the Blue Ridge Parkway Archives.)

REAR OF THE PUCKETT CABIN IN 1939. This angle shows the chimney on the Puckett Cabin, which was laid by hand during construction before the Civil War. Stones were gathered from around the property and strategically placed into the cabin. This property was purchased in 1939, and Orlean Puckett was given only 30 days to vacate. (Courtesy of the Blue Ridge Parkway Archives.)

CLOSE-UP VIEW OF PUCKETT CABIN. This photograph, taken in September 1953, shows the hand-hewn logs of the Puckett Cabin after restoration. Purchased in 1939, the cabin was completely disassembled and then reassembled. Renovations in 1976 included a new shaker shingle roof. When it was first purchased, a tin roof was placed on the cabin. (Courtesy of the Blue Ridge Parkway Archives.)

MILEPOST 189.8, PUCKETT CABIN. On July 18, 1960, this photograph was taken of a visitor and interpretation board at Puckett Cabin, a popular stopping point on the Blue Ridge Parkway in Carroll County, Virginia. Only a foot path led to the cabin in 1960. Today a parking lot, a split-rail fence, and other interpretive signs guide the visitor through Orlean Puckett's life. (Courtesy of the Blue Ridge Parkway Archives.)

Stone Mountain "Balcony." In the 1940s, interpretive signs were placed at each overlook, naming the overlook and stating information about the scene. Local residents often called them "balconies." Overlooks were carefully chosen during the initial surveys. This photograph was taken sometime in the 1950s at the Stone Mountain Overlook in Alleghany County, North Carolina. (Courtesy of a private collection.)

Family in Western North Carolina. This appears to be the primary type of overlook with a stunning vista. In the late 1950s, vacations along the Blue Ridge Parkway included many beautiful family photographs with scenery behind. Retaining walls were built during the early stages of parkway construction. Covered with local stone, the walls were to blend into the scene. (Courtesy of a private collection.)

CHEROKEE VACATION. After more than five years of debate, in February 1940, the Ridge Route was accepted by the Cherokee Tribal Council as the path to be taken by the Blue Ridge Parkway. They received $40,000 for the right-of-way through the reservation, with hopes of a start to ending their economic depression. (Courtesy of a private collection.)

CHEROKEE CHIEF. The Cherokees opened up their reservation to the public in hopes of economic progress. They were encouraged by the Bureau of Indian Affairs after many years of oppression to demonstrate their crafts and share their heritage. In the 1950s, with the onset of television, interest grew in the Native American culture. (Courtesy of a private collection.)

SUNSHINE TEEPEE. Exploring teepees became a favorite pastime of travelers in the 1950s. With the expansion of television and the Western genre, Americans became very interested in the Native American culture. Original teepees were assembled with poles and tanned leather. (Courtesy of a private collection.)

PICNIC AREA IN 1953. Comfort stations and picnic areas (similar to the one pictured below) were built in the early stages of the construction by the CCC. They would have landscaped the picnic area and built the facilities and tables. A contractor would have paved this winding road. (Courtesy of a private collection.)

MOSES CONE HOUSE IN 1953. Bertha Cone, widow of Moses Cone, donated her late husband's estate, Flat Top Manor, to the state upon her death in 1947. The state deeded the property to Blue Ridge Parkway. Today it houses the Southern Highlands Crafts Guild. (Courtesy of the Blue Ridge Parkway Archives.)

ROOT CELLAR RESTORATION IN 1953. Milepost 5.8 boasts the Humpback Rocks Center. Logs were hand cut by local artisans for the restoration. The 795 acres of Humpback Rocks recreation area are notable for the rock fence, reputedly built by slaves of a plantation owner, which separates the gap and Greenstone Overlook. During summer months, demonstrations of folk crafts are held in the buildings. (Courtesy of the Blue Ridge Parkway Archives.)

ROOT CELLAR IN 1953. The root cellar was a necessary underground room suitable for storage of consumable goods. Until the advent of electrical refrigeration, the root cellar was the most predominant means of storage of vegetables and meats. Wine cellars are a specialized type of root cellar in which the goal is consistent or reduced ambient temperature. (Courtesy of the Blue Ridge Parkway Archives.)

SORGHUM MOLASSES DEMONSTRATION IN 1969. Fall brings the tradition of preparing molasses. Molasses is a dark brown, thick syrup by-product from the processing of the sorghum cane. The quality of molasses depends on the maturity of sorghum, the amount of juice extracted, and the method of extraction. (Courtesy of the Blue Ridge Parkway Archives.)

HAPPY CHIEF. Known as the friendliest tribe in America, the Cherokees offered to have their photographs taken with visitors as a vacation memento. Two Cherokee, North Carolina, principal attractions for roadsiders are its bear pits and Chief Henry, the world's most photographed Native American. Chief Henry (Henry Lambert) has appeared on at least 25 postcards over the past 50 years. In 1998, Henry had his golden anniversary posing. Vacationers expect to see a Native American in feathers, but the Cherokee only wore feathers in ceremonies. Cherokees never wore huge feather headdresses like the Woodland or Plains people and only wore a feather in time of war or during a ball game similar to lacrosse—and then it was just a single feather. The women used white turkey feathers when making a tear dress. (Courtesy of a private collection.)

Seven

THE MISSING LINK

In 1966, ninety-five percent of the Blue Ridge Parkway was complete, except for a 7.7-mile "missing link" around privately owned Grandfather Mountain in North Carolina. Several unsuccessful attempts were made to acquire the right-of-way, but the owner, Hugh Morton, vehemently objected to the destructive proposals made for getting over and through the mountain.

Finally, in the 1970s, an agreement was made between Morton and the National Park Service. This agreement included a revolutionary viaduct that would float around the mountain. Minimal damage would be done to Grandfather Mountain.

A viaduct is described as a bridge designed in segments that carry the road high above ravines or across the shoulders of mountains.

This chapter has photographs of the road leading to nowhere before the construction of Linn Cove Viaduct began in 1979. It was designed by Figg and Muller Engineers, Inc., of Tallahassee, Florida.

Exactly 153 segments were precast at a facility near the parkway, transported to the site, and assembled on-site using continuous construction methods. A custom crane was used to lower each segment into place. The only on-site work done was predrilling for the foundation of the seven support footings. Only the Blue Ridge Parkway can access this bridge. No other access roads exist.

Today the Linn Cove Viaduct at milepost 304 has won at least 11 design awards. The S-curved bridge contains 180-foot spans of alignment geometry used in road construction, including reverse curvature. Linn Cove Viaduct is 4,400 feet above sea level.

To the south of the bridge, the Blue Ridge Parkway maintains a visitor center with a trail leading to the bridge.

Jasper Construction Company completed the construction in 1983 at a cost of $10 million. The entire 469-mile route of the Blue Ridge Parkway was completed four years later and was opened to the public for travel.

Surprisingly, there are two other viaducts on the parkway. One is near Laurel Fork, Virginia, and one is near Round Meadow at milepost 179, north of Laurel Fork.

AERIAL PHOTOGRAPH OF THE MISSING LINK. In August 1976, this aerial photograph shows the "missing link." Grandfather Mountain stands to the left in this photograph looking north on the Blue Ridge Parkway. Hugh Morton argued that it would destroy the mountain to dynamite and clear for the road or build a tunnel through the mountain. (Courtesy of the Blue Ridge Parkway Archives.)

LINVILLE INTERCHANGE AT MILEPOST 312. This aerial photograph of the Linville interchange shows the southern end of the missing link of the Blue Ridge Parkway. Highway 221 was the bypass between Blowing Rock and Linville, North Carolina, until the link was completed in 1987. (Courtesy of the Blue Ridge Parkway Archives.)

INSPECTION CREW. Planning and inspecting the path for the Linn Cove Viaduct required careful attention to detail by this group. From left to right are Gary Kleindinst; Bob Schreffler; Chief of Interpretation Bob Bruce; a federal highways employee; Gary Johnson, project inspector; and the pilot. On this day in August 1976, they were flying equipment into the site to do soil borings. (Courtesy of the Blue Ridge Parkway Archives.)

CONSTRUCTION CRANE. Construction cranes have telescoping, lattice, or articulating booms. This crane was used to lower sections of the Linn Cove Viaduct into position. As the bridge grew, crews would progress to the next section. Support placement was articulated precisely. (Courtesy of the Blue Ridge Parkway Archives.)

PRECAST SEGMENT. The 153 precast segments were assembled on-site using continuous construction methods, and a custom crane was used to lower each one into place. The only on-site work done was predrilling for the foundation of the seven support footings. (Courtesy of the Blue Ridge Parkway Archives.)

CANTILEVERED SEGMENTS. Carrying the road high above treetops and ravines, the segments were placed into position through a technique called progressive erection. Each segment was match casted to its adjoining segments. The viaduct was completed in 1983 after several years of construction. (Courtesy of the Blue Ridge Parkway Archives.)

THE END IN 1976. The parkway stopped near Grandfather Mountain until negotiations between Hugh Morton and the National Park Service were settled in the 1970s. In 1955, just as the park service was seizing lands by eminent domain, Hugh Morton was making business plans. His plans did not agree with the park plans, and he would not agree to sell. (Courtesy of the Blue Ridge Parkway Archives.)

MASONRY WORK. Masons cut stones on-site for the Linn Cove visitor center. As called for in the National Park Service regulations, local stones were used to face parts of the building and retaining wall. A jackhammer is used to split the rock into the desired shape. (Courtesy of the Blue Ridge Parkway Archives.)

VERTICAL VIEW OF VIADUCT. Double piers were built to support the Round Meadows Viaduct near milepost 179.4. In November 1938, with the leaves gone, a good view of the piers was photographed by one of the engineering staff. The southeast side of the viaduct shows the piers and steel used to support the concrete. (Courtesy of the Blue Ridge Parkway Archives.)

No. 3 PIER OF ROUND MEADOW VIADUCT. This angle shows the height of the viaduct at Round Meadow. In November 1938, construction appears to be almost complete, with some timbers lying on the hillside nearby. A high viaduct provides a smooth ride for the visitor. (Courtesy of the Blue Ridge Parkway Archives.)

SCAFFOLDING ON THE LAUREL FORK VIADUCT. Pier No. 4 is under construction in this photograph of the Laurel Fork Viaduct, taken in July 1937. Each section of the pier was formed with timbers and then poured on top of the section below. Concrete was mixed on-site and hauled up to the area that needed it. (Courtesy of the Blue Ridge Parkway Archives.)

PANORAMA OF LAUREL FORK VIADUCT. This panorama of the Laurel Fork Viaduct displays the irony of progress and the past season's haystacks, which were hand built. The haystacks were towered over by the large modern bridge. Ashe County, North Carolina, was mostly a farming community in the 1930s. Three piers, with arched centers, were built for this viaduct. (Courtesy of the Blue Ridge Parkway Archives.)

LAUREL FORK VIADUCT. Stationed at milepost 248.9 is the Laurel Fork Viaduct in Ashe County, North Carolina. Not as well known as Linn Cove but just as important, this viaduct carries the road high above beautiful farmland without disrupting nature. Constructed in 1937, this is part of one of the first completed sections of the parkway. (Courtesy of the Blue Ridge Parkway Archives.)

Eight

CRAFT DEMONSTRATORS

Mountain folk learned to make a living by sharing their crafts with the public. This chapter contains photographs archived with the Blue Ridge Parkway headquarters in Asheville, North Carolina. They showcase many handcrafts done by people living in the Blue Ridge Mountains and were taken by various photographers documenting the craft skills.

Some of the crafts shown in this chapter are fiber weaving, chair caning, basket weaving, whittling, apple butter making, sorghum molasses preparation, and the construction of musical instruments.

Most everything was made by hand in the mountains, including dwellings. Log cabins were hewn from logs found on the property. Rocks were gathered for foundations. Homemade cement was mixed from the sand found in the flowing rivers. Food was grown by the local farmer for his family, while extra was sold at market or to neighbors. Clothes were made from handspun fibers grown on the farm.

The Blue Ridge crafts heritage was founded in North Carolina and Virginia. Visitors from all over the world travel to the Blue Ridge Mountains in search of fine crafts. Here visitors can see artists at work in their studios, participate in hands-on demonstrations, view a great variety of crafts at galleries and museums, and learn about the crafts heritage. Today many people come to the region's venerable craft schools—for a weekend, a week, or longer—to learn a craft, refine their skills, or learn about Appalachian arts and crafts.

During the Depression, the federal Works Progress Administration fostered crafts projects as well as public works and murals. This enabled crafts to flourish at a local level. At the same time, American art programs began to include craft studies in their curricula.

VIADUCT STONEMASON.
Throughout the Blue Ridge Parkway
project, local stone was used to
face bridges and tunnels. This
stonemason was creating the facing
to be used at the Linn Cove Viaduct
next to Grandfather Mountain.
Known for their talent, stonemasons
from Italy were hired in the early
construction stages of the parkway.
(Courtesy of the Blue Ridge Parkway
Archives.)

GUN STOCK MANUFACTURING. On September 2, 1959, Reed Clifton demonstrated the art of
gun stock manufacturing in Vesta, Virginia. Rifles and shotguns bore the manufacturer's signature
in the early days of America. High-quality wood was used to carve elaborate scenes or details.
Custom-fitted stocks ensured accurate shooting by the serious shooter. (Courtesy of the Blue
Ridge Parkway Archives.)

MABRY MILL BLACKSMITH. This blacksmith is creating objects from iron or steel by "forging" the metal, which means using hand tools to hammer, bend, cut, and shape metal in its non-liquid form. Usually the metal is heated until it glows red or orange as part of the forging process. (Courtesy of the Blue Ridge Parkway Archives.)

WHITTLING. The art of carving shapes out of raw wood with a knife is called "whittling." Whittling was a big hobby for country folks in the late 1800s and early 1900s. Typically a pocketknife is the instrument of choice, and oftentimes figurines are the end result. (Courtesy of the Blue Ridge Parkway Archives.)

HANDMADE CHAIR. Handmade chairs were very common in the early 1900s. Very few tools were needed to craft a simple chair. Often one-of-a-kind chairs were a work of art made to fit the person receiving the chair. This chair was made in 1953 at a demonstration near Fancy Gap, Virginia. (Courtesy of the Blue Ridge Parkway Archives.)

JIM SOCKWELL POTTERY MAKING. History and memories are spun by Appalachian potters. Early pieces of pottery included milk jugs, crocks, pitchers, and decorative vases. Cherokee pottery served not only in day-to-day uses but also ceremonial uses. Today it is considered fine art. (Courtesy of the Blue Ridge Parkway Archives.)

Simon P. Scott Tanning Hides. Scott demonstrates the tanning of hides in 1941 near Meadows of Dan, Virginia. Here he is scraping the hide after removal from the tanning vat and preparing the hide for oiling. Oiling waterproofed the hides to be used as hats, coats, and tents. The term "tanning" refers to the concept of infusing the animal hide with the preservative "tannic acid." This prevents the skin from rotting. The most common method, historically, was "brain tanning." There is tannic acid, as well as oils and conditioners, in the brain that will transform a raw piece of animal hide into supple, garment-grade leather. The hide is removed from the carcass and then laced to a rigid frame. The hair and dermis is scraped from the outer layer of skin, and the fatty sub-tissue is scraped from the underside. (Courtesy of the Blue Ridge Parkway Archives.)

SPINNING AT BRINEGAR CABIN. This century-old spinning wheel is still used in craft demonstrations in the summer months at Brinegar Cabin at milepost 238.5. Other demonstrations may include quilting, old-time mountain music, weaving, and other crafts unique to the beginning of the 20th century. (Courtesy of the Blue Ridge Parkway Archives.)

TURNING A CHAIR LEG. Arvel Woody and his brother Walter entered the chair-making business after World War II. While the current designs have been modernized, the mortises are still cut with a machine manufactured in 1893, and the chairs continue to be assembled without glue. (Courtesy of a private collection.)

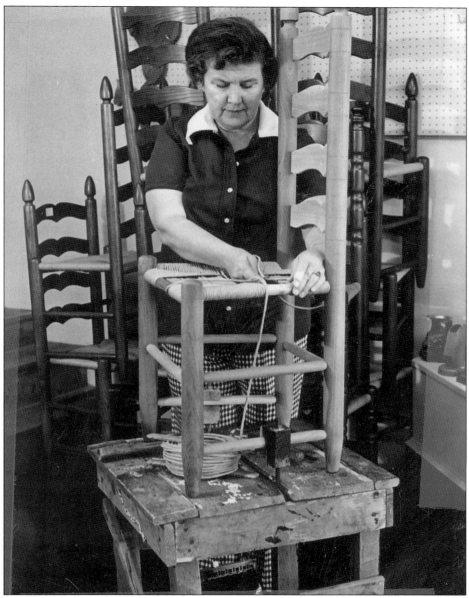

SEAT WEAVING AT WOODY'S CHAIR SHOP. Splints, sometimes referred to as "splits," are prepared strips of ash, oak, reed, or hickory bark woven around the seat rungs or dowels of chairs and rockers, usually in a herringbone-twill or basket-weave design. The rustic or lattice weave uses rawhide strips that are woven on chairs, rockers, and couches in a very open weave. Arvel Woody's grandfather, Arthur, made traditional "mule-eared" chairs. They were taken by ox-drawn sled to Marion and Forest City, North Carolina, to be exchanged for coffee and sugar at a rate of three chairs per $1. Arthur's career as a chairmaker coincided with the handicraft revival, and his work became popular in the North. He had an endless variety of chairs. Shipping chairs across the country to cities like Boston was common for him. When he and his son began teaching chair bottoming at Penland School in the 1920s, his reputation grew. (Courtesy of the Blue Ridge Parkway Archives.)

BIBLIOGRAPHY

Blouin, Nicole. *Waterfalls of the Blue Ridge: A Hiking Guide to the Cascades of the Blue Ridge Mountains*. Birmingham, AL: Menasha Ridge Press, 2003.

Davids, Richards C. *The Man Who Moved A Mountain*. Philadelphia: Fortress Press, 1970.

Jolley, Harley. *The Blue Ridge Parkway*. Knoxville, TN: University of Tennessee Press, 1969.

Welhem, E. J. *The Blue Ridge; Man and Nature in Shenandoah National Park and Blue Ridge Parkway*. Charlottesville, VA: University of Virginia Press, 1968.

Whisnant, Anne Mitchell. *Super-scenic Motorway: A Blue Ridge Parkway History*. Chapel Hill, NC: University of North Carolina Press, 2006.

National Park Service, Blue Ridge Parkway: www.nps.gov.blri

National Park Service, Shenandoah National Park: www.nps.gov/shen

INDEX

DISCOVER THOUSANDS OF LOCAL HISTORY BOOKS FEATURING MILLIONS OF VINTAGE IMAGES

Arcadia Publishing, the leading local history publisher in the United States, is committed to making history accessible and meaningful through publishing books that celebrate and preserve the heritage of America's people and places.

Find more books like this at
www.arcadiapublishing.com

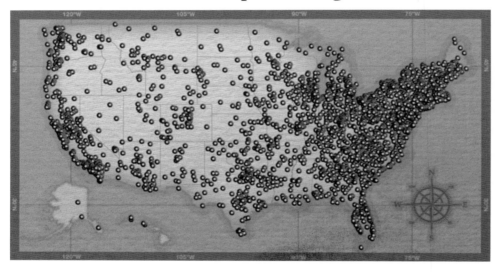

Search for your hometown history, your old stomping grounds, and even your favorite sports team.

Consistent with our mission to preserve history on a local level, this book was printed in South Carolina on American-made paper and manufactured entirely in the United States. Products carrying the accredited Forest Stewardship Council (FSC) label are printed on 100 percent FSC-certified paper.

MADE IN THE USA